Science in the News
Critical Thinking Worksheets

Eye on the Environment

HOLT, RINEHART AND WINSTON
A Harcourt Classroom Education Company

Austin • New York • Orlando • Atlanta • San Francisco • Boston • Dallas • Toronto • London

To the Teacher

The video segments in the **CNN Presents Science in the News: Eye on the Environment** program bring the knowledge and experience of the CNN news team right into your classroom. Show students the relevance of environmental science to everyday life with interesting news segments that relate to the topics they are studying in class. In this video, students will see the efforts of biologists, environmentalists, students, governments, and many others who are using science to answer important environmental questions. The segments selected for this video will help students learn more about the role of plants, animals, and Earth's processes in shaping the environment. Students will be encouraged to see the influence of technology on the environment and will learn how environmental issues are addressed on the local, national, and international level.

The Critical Thinking Worksheets contain thought-provoking questions that encourage students to carefully examine the information presented in each segment. Students can sharpen their listening and critical thinking skills as they are challenged to evaluate each segment. The worksheets also help to focus students' attention on the topics in each segment. You may wish to have students use their textbook as a resource for answering questions. In addition, topics covered in the segment and worksheets can be used as the basis for student-led discussions. The worksheet questions are also available in the While Viewing sections of the Teacher's Guide that accompanies the *Eye on the Environment* video. The Teacher's Guide also contains discussion points, research ideas, and other information that will help you and your students get the most from these worksheets.

Copyright © by Holt, Rinehart and Winston

All rights reserved. No part of this publication may be reproduced or transmitted in any form or by any means, electronic or mechanical, including photocopy, recording, or any information storage and retrieval system, without permission in writing from the publisher.

Teachers may photocopy complete pages in sufficient quantities for classroom use only and not for resale.

Art and Photo Credits
All work, unless otherwise noted, contributed by Holt, Rinehart and Winston.
Front cover, Jeffry Myers/Panoramic Images

Printed in the United States of America

ISBN 0-03-056553-7

3 4 5 6 085 03 02 01 00

Contents

	Segment Title	Page Number
Segment 1	Hazy Days	1
Segment 2	Development and Preservation	2
Segment 3	Hummingbird Mites	3
Segment 4	Monarch Migrations	4
Segment 5	Eagles and DDT	5
Segment 6	Biosphere Pioneers	6
Segment 7	Vanishing Rain Forest	7
Segment 8	Shrinking Wetlands	8
Segment 9	Watch for Flooding	9
Segment 10	China's Superdam	10
Segment 11	Smog Problems in Mexico	11
Segment 12	Greening Sudbury	12
Segment 13	Global Warming	13
Segment 14	CO_2 and Arctic Ozone	14
Segment 15	Tropical Reforestation	15
Segment 16	Prairie Restoration	16
Segment 17	Fish Farming	17
Segment 18	The Fire Ants and the Fungus	18
Segment 19	What's Slithering in Guam?	19
Segment 20	The Brink of Extinction	20
Segment 21	Fusion for Power?	21
Segment 22	Geothermal Energy	22
Segment 23	Talking Trash	23
Segment 24	Treating Toxic Waste	24
Segment 25	Critical Masses	25
Segment 26	Declining Birthrates	26
Segment 27	A Climate Conference	27
Segment 28	Something Worth Saving	28
Answer Key		**29**

Name _____ Date _____ Class _____

Science in the News: Critical Thinking Worksheets

Segment 1
Hazy Days

1. What was the primary cause of the fires in Mexico?

   _____
   _____
   _____
   _____

2. Name two types of pollution other than air pollution that can cross international borders.

   _____
   _____
   _____
   _____

3. What is a challenge related to solving international environmental problems like the one caused by Mexico's fires?

   _____
   _____
   _____
   _____

4. Name two ways science can help solve the air pollution problems in Mexico.

   _____
   _____
   _____
   _____
   _____
   _____

Name_____ Date_____ Class_____

Science in the News: Critical Thinking Worksheets

Segment 2
Development and Preservation

1. Why did it take so long for the participants in the development discussion to form an agreement?

   _____
   _____
   _____
   _____

2. Identify three steps involved in determining how land would best be used.

   _____
   _____
   _____
   _____
   _____

3. What is the developer referring to when he mentions a "no surprises" plan?

   _____
   _____
   _____
   _____
   _____

4. Why are some environmentalists dissatisfied by the compromise made between developers and the government?

   _____
   _____
   _____
   _____
   _____

Name_____ Date_____ Class_____

Science in the News: Critical Thinking Worksheets

Segment 3
Hummingbird Mites

1. Identify the type of interaction between the mites, flowers, and hummingbirds.

   _____
   _____
   _____
   _____
   _____

2. Formulate a hypothesis about another possible relationship between the mites and the flowers.

   _____
   _____
   _____
   _____
   _____

3. Why would a biologist be interested in this symbiotic relationship?

   _____
   _____
   _____
   _____
   _____

4. What are some of the possible uses of this biological research?

   _____
   _____
   _____
   _____
   _____

Name_____ Date_____ Class_____

Science in the News: Critical Thinking Worksheets

Segment 4
Monarch Migrations

1. Describe some of the methods, tools, and data used in the monarch migration research.

   _____
   _____
   _____
   _____
   _____

2. What are two unanswered questions about monarch migration that scientists hope to answer with their research?

   _____
   _____
   _____
   _____

3. a. Name one possible factor that may cause monarch roosts to be endangered.

   _____
   _____
   _____

   b. How could the results of this research help solve this problem?

   _____
   _____
   _____

4. Can the methods used to study monarchs be used to study other migrating species? Explain your answer.

   _____
   _____
   _____
   _____

Name_____ Date_____ Class_____

Science in the News: Critical Thinking Worksheets

Segment 5
Eagles and DDT

1. **a.** Identify one way the artificial incubation and release program was successful.

   _____
   _____

   b. Identify one way the artificial incubation and release program was a failure.

   _____
   _____

2. Is DDT poisoning a greater threat to bald eagles and other vulnerable species than the threat of habitat destruction? Explain your answer.

   _____
   _____
   _____
   _____

3. What forms of pollution threaten both wildlife and humans? Give two examples.

   _____
   _____
   _____

4. Why do scientists return juvenile eagles to Catalina Island when they know the birds will eat DDT-contaminated fish?

   _____
   _____
   _____
   _____
   _____

Name _____ Date _____ Class _____

Science in the News: Critical Thinking Worksheets

Segment 6
Biosphere Pioneers

1. On what basis were ecosystem types chosen for Biosphere 2?

   _____
   _____
   _____
   _____
   _____

2. Formulate a hypothesis about the diet of the scientists living in Biosphere 2.

   _____
   _____
   _____
   _____
   _____

3. How do you think Biosphere 2 inhabitants dispose of their waste products?

   _____
   _____
   _____

4. In the years following this news report, Biosphere scientists had serious difficulties producing sufficient oxygen because microbes in the soil consumed much of the air supply. Why might scientists have hesitated to remove the microbes from the soil?

   _____
   _____
   _____
   _____
   _____

Name_____ Date_____ Class_____

Science in the News: Critical Thinking Worksheets

Segment 7

Vanishing Rain Forest

1. Name two competing demands for the resources found in rain forests.

   _____
   _____
   _____
   _____
   _____

2. How do the traditional slash-and-burn technique and logging operations affect the biodiversity of Nicaragua's tropical forests?

   _____
   _____
   _____
   _____

3. Can people relocate to the forests without harming the environment? Explain your answer.

   _____
   _____
   _____
   _____
   _____

4. What is the difference between preservation and conservation?

   _____
   _____
   _____
   _____

Name _____ Date _____ Class _____

Science in the News: Critical Thinking Worksheets

Segment 8
Shrinking Wetlands

1. Give three examples of the benefits of wetlands.

   _____
   _____
   _____
   _____

2. How does the loss of wetlands affect fisheries and local economies?

   _____
   _____
   _____
   _____

3. Name two incentives that the government could offer to farmers that might encourage them to preserve wetlands on their property.

   _____
   _____
   _____
   _____

4. Do you believe artificially constructed wetlands could benefit humans and the environment as effectively as natural wetland ecosystems? Explain your answer.

   _____
   _____
   _____
   _____

Name_____ Date_____ Class_____

Science in the News: Critical Thinking Worksheets

Segment 9
Watch for Flooding

1. Name two reasons why people commonly choose to build and live on a flood plain.

   _____
   _____
   _____
   _____
   _____

2. Give two examples of ways flooding can be prevented in developed areas.

   _____
   _____
   _____
   _____
   _____
   _____

3. How can flooding contribute to ground and surface water pollution?

   _____
   _____
   _____
   _____

4. What does the speaker in the video mean when he says that "nature wins"?

   _____
   _____
   _____
   _____
   _____

Name_____ Date_____ Class_____

Science in the News: Critical Thinking Worksheets

Segment 10
China's Superdam

1. How will the Three Gorges Dam contribute to the rapid growth of cities along the banks of the Yangtze River?

   _____
   _____
   _____
   _____
   _____

2. How might the dam contribute to pollution of the Yangtze River?

   _____
   _____
   _____

3. Besides water pollution, what is another possible environmental effect of building the dam?

   _____
   _____
   _____

4. Describe two economic, cultural, or environmental trade-offs involved in the Three Gorges Dam project.

   _____
   _____
   _____
   _____
   _____
   _____

Name_____ Date_____ Class_____

Science in the News: Critical Thinking Worksheets

Segment 11
Smog Problems in Mexico

1. Identify the natural phenomenon discussed in the video that can aggravate existing air pollution.

   _____
   _____
   _____
   _____

2. Name the two greatest sources of pollutants in Mexico City and other industrial cities.

   _____
   _____
   _____
   _____

3. How might people in your community react to a limit on the number of days per week they were allowed to use their cars? Explain your answer.

   _____
   _____
   _____
   _____

4. What is the biggest trade-off involved in shutting down factories to lessen pollution?

   _____
   _____
   _____
   _____
   _____

Name_____ Date_____ Class_____

Science in the News: Critical Thinking Worksheets

Segment 12
Greening Sudbury

1. How did the emissions from the mining industry result in such devastating pollution?

_____
_____
_____
_____

2. Why did a mining company build a taller smokestack before it implemented more-effective pollution control measures?

_____
_____
_____
_____

3. What measures have been taken to clean up the local environment?

_____
_____
_____
_____

4. a. How have mining operations and regulations changed since 1972?

_____
_____

b. Identify one of the reasons for these changes.

_____
_____
_____

Name_____ Date_____ Class_____

Science in the News: Critical Thinking Worksheets

Segment 13
Global Warming

1. Identify a particular ecosystem, and describe how global warming might affect this ecosystem.

   _____
   _____
   _____
   _____

2. Do you think reducing industrial and automotive emissions is the most effective way to stop global warming? Explain your answer.

   _____
   _____
   _____
   _____
   _____

3. How would changes in the Earth's seasons disrupt agriculture and food supplies?

   _____
   _____
   _____
   _____

4. The recommendations of some scientists to control global warming will require a great deal of work and money. Do you think the cost of these efforts is justified? Explain your answer.

   _____
   _____
   _____
   _____
   _____

Name_____ Date_____ Class_____

Science in the News: Critical Thinking Worksheets

Segment 14
CO_2 and Arctic Ozone

1. How do global warming gases, such as CO_2, increase stratospheric ozone depletion?

   _____
   _____
   _____
   _____
   _____

2. Identify at least two current attempts to slow CO_2 emissions.

   _____
   _____
   _____
   _____

3. How would the carbon tax mentioned in the video affect the economy and the environment?

   _____
   _____
   _____
   _____
   _____

4. Name one major limitation of using computer models to predict changes in Earth's climate.

   _____
   _____
   _____
   _____

Name_____ Date_____ Class_____

Science in the News: Critical Thinking Worksheets

Segment 15
Tropical Reforestation

1. Why is the Costa Rican reforestation project described in the video privately funded?

   _____
   _____
   _____
   _____

2. How do researchers determine which trees are best suited for reforestation of a particular area?

   _____
   _____
   _____
   _____

3. Identify two benefits of reforestation projects and of planting trees in general.

   _____
   _____
   _____
   _____

4. Name two reasons why certain types of land use, such as clearing forests for grazing, may be unsuitable for a particular ecosystem.

   _____
   _____
   _____
   _____
   _____

Name _____ Date _____ Class _____

Science in the News: Critical Thinking Worksheets

Segment 16
Prairie Restoration

1. Identify two ways a suburb could benefit from preserving prairie land.

   _____

   _____

   _____

   _____

2. **a.** What are some threatened native plants or animals in your area?

   _____

   _____

   b. Are there any ways threatened plant or animal populations in your area can be increased?

   _____

   _____

   _____

   _____

3. Describe one possible benefit of saving a threatened or endangered species' genes.

   _____

   _____

   _____

4. Give three reasons why a valid "use" of land is to allow it to return to its natural state.

   _____

   _____

   _____

   _____

Name _____ Date _____ Class _____

Science in the News: Critical Thinking Worksheets

Segment 17
Fish Farming

1. Name two species besides those mentioned in the video that could be successfully raised using aquaculture. Explain why you chose these species.

   _____
   _____
   _____
   _____

2. **a.** Name one benefit of using aquaculture as a major food source.

   _____
   _____
   _____

   b. Name one drawback of using aquaculture as a major food source.

   _____
   _____
   _____

3. Why are some people concerned about the food safety of fish caught in the wild?

   _____
   _____
   _____
   _____

4. What does the phrase "building a better fish" mean?

   _____
   _____
   _____

Name _____ Date _____ Class _____

Science in the News: Critical Thinking Worksheets

Segment 18

The Fire Ants and the Fungus

1. Name two ways that exotic species may be accidentally introduced to a new ecosystem.

   _____
   _____
   _____
   _____

2. Why is the source country often the same for both the pest and the possible biological solution?

   _____
   _____
   _____
   _____

3. Why did the researchers patent the information they learned about the fungus?

   _____
   _____
   _____
   _____

4. Describe two trade-offs between the harm the fire ants cause and the effects of pest-control agents on the environment and human health.

   _____
   _____
   _____
   _____
   _____

Name _____ Date _____ Class _____

Science in the News: Critical Thinking Worksheets

Segment 19

What's Slithering in Guam?

1. Why don't officials introduce a predator species to control the nonnative snake population?

   _____
   _____
   _____

2. Why is it important to save the last breeding pair of a species?

   _____
   _____
   _____
   _____

3. Identify the primary reason that many species around the world need captive-breeding programs to ensure their survival.

   _____
   _____
   _____
   _____
   _____
   _____

4. Compare the cost of developing and maintaining a captive-breeding program with the cost of a species' extinction.

   _____
   _____
   _____
   _____
   _____
   _____

Name _____ Date _____ Class _____

Science in the News: Critical Thinking Worksheets

Segment 20
The Brink of Extinction

1. Explain why some environmentalists like the mixed land-use plan and others do not.

   _____
   _____
   _____
   _____

2. Why do you think some species' populations reach dangerously low levels before a problem is recognized?

   _____
   _____
   _____
   _____
   _____

3. In your opinion, will compromises between developers and environmentalists be effective in saving endangered species? Explain your answer.

   _____
   _____
   _____
   _____

4. Why do you think the federal government favors the "whole ecosystem" approach of habitat and species preservation?

   _____
   _____
   _____
   _____

Science in the News: Critical Thinking Worksheets

Segment 21

Fusion for Power?

1. In order to generate temperatures of 400 million degrees Celsius, scientists had to use a lot of energy and money. How can the high costs of this research be justified?

2. Why do you think fusion scientists want to keep the atomic particles from touching the walls of the tube?

3. Why is fusion power called a "clean" industry?

4. Why is fusion referred to as a limitless energy source?

Name _____ Date _____ Class _____

Science in the News: Critical Thinking Worksheets

Segment 22
Geothermal Energy

1. Identify the primary cost-efficient aspect of a geothermal energy system.

   _____
   _____
   _____
   _____

2. Is a backup energy source needed to supplement the geothermal energy source? Why or why not?

   _____
   _____
   _____
   _____
   _____

3. What are the environmentally friendly aspects of geothermal energy?

   _____
   _____
   _____
   _____
   _____

4. Identify two limitations of geothermal energy as an energy source.

   _____
   _____
   _____
   _____
   _____

Name_____ Date_____ Class_____

Science in the News: Critical Thinking Worksheets

Segment 23
Talking Trash

1. List everything that you threw out over the last 3 days. How much of it do you think could have been reused or recycled?

   _____
   _____
   _____
   _____
   _____

2. Why don't all communities have a recycling program?

   _____
   _____
   _____
   _____
   _____

3. Describe the "life" of a single product you use every day.

   _____
   _____
   _____
   _____
   _____

4. Garbage items, such as a banana peel, a pair of leather shoes, or a book, take a long time to biodegrade. How could these items be reused or recycled instead of thrown away?

   _____
   _____
   _____

Name_____ Date_____ Class_____

Science in the News: Critical Thinking Worksheets

Segment 24
Treating Toxic Waste

1. Why is the process of converting chemical wastes back to their simplest form beneficial?

   _____
   _____
   _____
   _____

2. Name one kind of toxic waste that has not yet been tested with molten metal technology.

   _____
   _____
   _____
   _____

3. Why is the Department of Energy willing to fund the development of molten metal technology?

   _____
   _____
   _____
   _____

4. What are the economic advantages of this technology over existing hazardous-waste treatment methods?

   _____
   _____
   _____
   _____
   _____

Science in the News: Critical Thinking Worksheets

Segment 25

Critical Masses

1. How does living in extremely overcrowded cities affect humans' average life expectancy?

2. Describe the relationships between population, poverty, and pollution.

3. Why do people move to overcrowded cities even though many of these cities have more sickness, homelessness, poverty, and crime?

4. Is rapid population growth occurring mainly in cities of developed countries or developing countries?

Name _____ Date _____ Class _____

Science in the News: Critical Thinking Worksheets

Segment 26
Declining Birthrates

1. Identify two reasons for the steady population decline in Japan.

   _____
   _____
   _____
   _____

2. Why might the Japanese government be concerned about the declining population?

   _____
   _____
   _____
   _____
   _____

3. What could the Japanese government do to slow or halt Japan's population decline?

   _____
   _____
   _____
   _____
   _____

4. Why might people in rich, industrialized countries have fewer children than those in poorer, developing countries?

   _____
   _____
   _____
   _____
   _____

Name_____ Date_____ Class_____

Science in the News: Critical Thinking Worksheets

Segment 27
A Climate Conference

1. What does the attendee mean when he says that many countries lack real commitment to making changes to their environmental policies?

   _____
   _____
   _____
   _____

2. Identify one opinion voiced in the video that could be considered extreme.

   _____
   _____
   _____
   _____

3. Name one reason why the goals and needs of rich, industrialized countries might differ from those of poorer, developing countries.

   _____
   _____
   _____
   _____

4. Name an advantage and a disadvantage of a binding environmental agreement between countries.

   _____
   _____
   _____
   _____

Name_____ Date_____ Class_____

Science in the News: Critical Thinking Worksheets

Segment 28
Something Worth Saving

1. Why does the speaker in the video refer to the tropical forest as a "living laboratory"?

   _____
   _____
   _____

2. Name two possible properties of plants discovered in tropical rain forests that could be useful to humans.

   _____
   _____
   _____
   _____

3. Give an example of something besides plant samples that a "biological prospector" might be interested in collecting.

   _____
   _____
   _____

4. a. Name one way that preserving the rain forest can contribute to a sustainable world.

   _____
   _____
   _____

   b. How do scientists contribute to this effort?

   _____
   _____
   _____

Science in the News: Critical Thinking Worksheets   Eye on the Environment

Science in the News: Critical Thinking Worksheets Answer Key

Segment 1
Hazy Days
1. The primary cause of Mexico's fires was the severe drought left by El Niño.
2. Answers will vary. Sample answer: All types of pollution can cross international borders, including water pollution and light pollution.
3. Answers will vary. Sample answer: A challenge related to solving international environmental problems is helping countries overcome financial and cultural differences in order to work together.
4. Answers will vary. Sample answer: Science can be used to conduct thorough studies of the causes and effects of air pollution on the environment. Science can then be used to find ways to reduce those effects.

Segment 2
Development and Preservation
1. Answers will vary. Sample answer: The process took so much time because each side had a different idea of how the land should be managed. Also, environmental studies may take a long time to complete.
2. Answers will vary. Sample answer: Three steps are as follows: conducting biological studies and surveys; reviewing federal laws and regulations; and forming an agreement that satisfies parties on both sides of the issue.
3. A "no surprises" plan is one that assures developers and landowners that they will not lose building permits in the future due to environmental concerns.
4. Answers will vary. Sample answer: Some environmentalists are not satisfied because the agreement allows further development in an endangered ecosystem.

Segment 3
Hummingbird Mites
1. The interactions are mutually beneficial. The mites use the hummingbirds as transportation, the flowers use the birds as carriers of pollen to fertilize other plants, and the plants supply food and a home for the mites.
2. Answers will vary. Sample answer: The mites might use the flowers as a defense against predators by hiding within the plants. The mites might benefit the flower by preying on other insects that harm the plant.
3. Answers will vary. Sample answer: A biologist would be interested because the relationship between animals and plants is an important factor in the evolution of species.
4. Answers will vary. Sample answer: This research could help biologists learn about pest management for other plant species. Also, the research may offer clues to the nature of other relationships between birds, plants, and insects.

Science in the News: Critical Thinking Worksheets Answer Key

Segment 4
Monarch Migrations
1. The methods used included catching and tagging butterflies. The students collected data about when the butterflies were caught, their sex, and which direction they flew when they were released. Their tools included compasses, observation logs, computers, and identification tags.
2. Scientists want to learn more about the timing of migration and about how fast the monarchs are able to move.
3. a. Answers will vary. Sample answer: Monarch roosts could be endangered by deforestation.
   b. Answers will vary. Sample answer: This research could help by supplying data that shows how the butterfly population is affected by damage to the roosts.
4. Answers will vary. Sample answer: Yes; these methods are useful for studying all types of migrating populations because all migrations involve movement, direction, and the cooperation of large numbers of animals.

Segment 5
Eagles and DDT
1. a. The eagle population on Catalina Island increased.
   b. Answers will vary. Sample answer: Bald eagles are still in danger because scientists have not found a way to prevent the eagles from eating DDT-contaminated food.
2. Answers will vary. Sample answer: No; DDT was not manufactured after 1972, and DDT generally affects a small area, whereas habitat destruction is an ongoing, worldwide threat.
3. Answers will vary. Sample answer: Water and air pollution threaten both wildlife and humans.
4. The only way scientists can create healthy and natural populations of bald eagles is to allow them to reproduce in the wild. Also, environmentalists believe that the only way to change environmental policies and people's behavior is to keep endangered species visible to the public.

Segment 6
Biosphere Pioneers
1. Ecosystems were chosen according to their usefulness to the inhabitants. For example, a prairie ecosystem produces food, and a swamp ecosystem recycles water.
2. Answers will vary. Sample answer: The diet of the scientists in Biosphere 2 is probably mostly plant-based because plants take up less space and fewer resources than animals raised for food. Also, the scientists' diet is probably very simple because of the limited number of ingredients available.
3. Answers will vary. Sample answer: Waste products must be reused or recycled in some way because Biosphere 2 is a closed environment.
4. Answers will vary. Sample answer: Scientists might have hesitated to remove the microbes because the microbes helped to enrich the soil. Also, insecticides might have been necessary to kill the microbes. These insecticides could have polluted the Biosphere environment.

Science in the News: Critical Thinking Worksheets Answer Key

Segment 7
Vanishing Rain Forest
1. One competing demand for forest resources is the local population's short-term need for wood and the money that can be earned from selling it. Another demand is the Earth's long-term environmental need for rain forests and the animals and plants they support.
2. When an area of forest is destroyed, a variety of species of animals living in the forest are also destroyed. This harms the biodiversity of the rain forest.
3. Answers will vary. Sample answer: Yes; but they would need to support themselves in a way that does not destroy the natural habitat. The traditional practice of slash-and-burn agriculture would have to be replaced by a less-destructive form of agriculture.
4. Answers will vary. Sample answer: Preservation is the complete protection of natural resources or areas. Conservation is the controlled and limited use of natural resources.

Segment 8
Shrinking Wetlands
1. Wetlands are beneficial to the environment because they provide flood control, filter fresh water, and provide an important habitat for many animal and plant species.
2. The loss of wetlands reduces the local fish population. This decline hurts the fishing industry and the local economy.
3. Answers will vary. Sample answer: The government could offer to help make the land farmers currently use more profitable in exchange for a promise to preserve wetlands. The government could also offer to lower taxes for farmers that preserve wetland areas.
4. Answers will vary. Sample answer: Artificially constructed wetlands could offer some of the benefits that a natural wetland does, such as flood control and recreation. However, they might not provide a complete habitat for animals because they would lack the biodiversity of a natural wetland.

Segment 9
Watch for Flooding
1. People often live on a flood plain because their families traditionally lived in the area. In addition, flood plains are agriculturally fertile areas.
2. Installing levee systems and building dams can prevent some flooding. Flooding can also be reduced by restricting the construction of channels for barge travel.
3. When agricultural areas flood, fertilizers and other chemicals run off and leak into ground and surface water.
4. The speaker means that despite our efforts to control the flooding of rivers with the use of dams and canals, eventually rivers will flood because flooding is a natural and powerful part of the river's processes.

Segment 10
China's Superdam
1. The Three Gorges Dam will turn some small, inland cities into cities with international ports. This will create many new jobs. As more work becomes available, more people from the country will come to the city to work.
2. When the river is dammed, existing pollution will collect near the cities that produce it.
3. Another environmental effect will be the destruction of many ecosystems in and around the river.
4. Answers will vary. Sample answer: The economy of China will be enhanced by hydroelectric power. Also, more trade for inland areas will be possible. However, the environment and the cultural heritage along the river may be severely damaged by the rapid population growth. Also, the dam might cause pollution for future generations.

Science in the News: Critical Thinking Worksheets Answer Key

Segment 11

Smog Problems in Mexico

1. A weather event called a thermal inversion can aggravate existing air pollution.
2. The two greatest sources of pollutants are cars and industry.
3. Answers will vary. Sample answer: People might be angry about a driving limit because it may be more difficult to get to work or school. They might also be glad because driving limits would reduce traffic and air pollution.
4. Answers will vary. Sample answer: Shutting down factories helps improve air quality. However, workers lose pay when the factories are closed. This has an effect on other local businesses because people have less money to spend.

Segment 12

Greening Sudbury

1. The mines released acidic gas into the air. The gas combined with oxygen in the atmosphere to create sulfur dioxide. Sulfur dioxide combined with water to produce acid rain, which killed plants and animals and damaged human health.
2. Taller smokestacks spread the sulfur dioxide gas over a larger area, creating lower emission readings, even though the same amount of waste was being released.
3. The mining industry and the government have supported the liming and reseeding of the local environment. Liming changes acidic soil to neutral soil.
4. a. Government regulations have become more strict since 1972. Mining companies must now control their emission levels.
   b. The mining industry has been more willing to reduce emissions because it has found a way to make a profit by selling acid that was once released into the atmosphere.

Segment 13

Global Warming

1. Answers will vary. Sample answer: A forest ecosystem could be affected by global warming. Trees might not survive rising temperatures. The loss of many species of trees over time could disrupt the food chain within the ecosystem.
2. Answers will vary. Sample answer: Reducing automotive emissions is one of the best ways to fight global warming because emissions from cars are a major cause of the increase in greenhouse gases.
3. Plants' production and reproduction cycles are tied to the Earth's seasons. Rising temperatures would change those cycles and would probably keep plants from producing the same amount of food as they do now.
4. Answers will vary. Sample answer: The amount of money and effort required is justified by the need to return the Earth's ecosystem to its natural state and to protect plant and animal species that are endangered by global warming.

Segment 14

CO_2 and Arctic Ozone

1. Global warming gases, such as CO_2, reduce temperatures in the stratosphere, which encourages the formation of polar stratospheric clouds. These clouds then further reduce stratospheric ozone.
2. Car pools and ozone-watch days are currently being used to decrease CO_2 emissions.
3. Answers will vary. Sample answer: A carbon tax would help to control global warming by discouraging people from buying carbon-based fuels. However, this tax would hurt and possibly destroy economies that depend heavily on carbon-based fuels.
4. One major limitation of using computer models is that new factors may be introduced into the environment that cause the model to need constant adjusting.

Science in the News: Critical Thinking Worksheets Answer Key

Segment 15
Tropical Reforestation
1. Answers will vary. Sample answer: The reforestation project helps the loggers, who are generally associated with private companies. Also, the government of Costa Rica might not have the money to fund this project.
2. Researchers determine which trees are best suited by studying how quickly each species grows in different soil types. They also determine which plants can be sold for profit in each particular area.
3. Answers will vary. Sample answer: Reforestation supports animal populations in an ecosystem by providing food and shelter. Also, forests recycle gases in the atmosphere and produce oxygen.
4. Answers will vary. Sample answer: Clearing forests for grazing destroys the habitat of climbing animals and many birds that depend on trees for protection from predators. Also, the clearing of forests promotes the erosion of topsoil, which could hurt local farmers.

Segment 16
Prairie Restoration
1. Answers will vary. Sample answer: Prairie lands require few fertilizers; therefore, the use of polluting chemicals would decrease. A suburb that preserves prairie land could also attract wildlife, such as birds and insects.
2. a. Answers will vary. Refer students to their textbook, the Internet, or a local library, if necessary.
   b. Answers will vary, depending on how scarce the species has become. Students should be able to determine the answer by contacting the nearest office of the Environmental Protection Agency.
3. Answers will vary. Sample answer: Saving a threatened species' genes could be beneficial because the genes might be used to help fight future diseases. Saving genes could also be helpful in identifying which species have become extinct.
4. Answers will vary. Sample answer: Allowing land to return to its natural state protects natural plants, provides habitats for many animals, and makes it possible for the land to be developed as public parkland.

Segment 17
Fish Farming
1. Answers will vary. Sample answer: Bass and oysters could be raised using aquaculture. These organisms are fairly small and would not need as much room to move around as bigger animals.
2. a. Answers will vary. Sample answer: A benefit of aquaculture is that fish and shellfish can be raised almost disease-free.
   b. Answers will vary. Sample answer: One drawback is the high cost of farm-raising fish. Also, some people think farm-raised fish do not taste as good as fish caught in the wild.
3. There are serious dangers associated with eating fish caught in the wild because it is often difficult to know whether the fishes' environment and food source is polluted.
4. "Building a better fish" means raising fish that are healthy, safe, and inexpensive.

Science in the News: Critical Thinking Worksheets Answer Key

Segment 18
The Fire Ants and the Fungus
1. Answers will vary. Sample answer: Exotic species can be transported with imported food. Exotic species can also be introduced when people import animals as pets.
2. The source is often the same because in the pest's natural environment, the pest population may have been controlled by another local species.
3. Answers will vary. Sample answer: The researchers patented their information so that they could prevent other people from selling their product as an insecticide.
4. Answers will vary. Sample answer: Eliminating the fire ants will save crops and farm animals. However, some pest-control agents may introduce other exotic species that could hurt the environment and harm people. Chemicals used to kill fire ants might also kill insects that are beneficial to humans.

Segment 19
What's Slithering in Guam?
1. Officials do not introduce another exotic species because it might destroy native species in addition to the nonnative snake population.
2. It is important because if the last breeding pair is not saved, the species will become extinct.
3. As with the local bird species in Guam, other species worldwide are becoming extinct in the wild due to human-made problems, such as habitat loss and environmental pollution. Captive-breeding programs can halt or slow the extinction of a species.
4. Answers will vary. Sample answer: The cost of a captive-breeding program is minor compared with the cost of extinction because once a species is extinct, that part of the ecosystem and food chain is permanently destroyed.

Segment 20
The Brink of Extinction
1. Answers will vary. Sample answer: Some environmentalists like the plan because it saves some of the birds' habitat. Others do not like the plan because it allows developers to continue to build in fragile ecosystems.
2. Answers will vary. Sample answer: A species' population might decrease very quickly, making the problem difficult to recognize before the species becomes threatened or extinct. Also, the reduction of a naturally small population might be difficult to notice.
3. Answers will vary. Sample answer: Compromises between developers and environmentalists will never be fully effective in saving endangered species because they will always allow some further destruction of natural habitat.
4. Answers will vary. Sample answer: The federal government favors the "whole ecosystem" approach because it focuses on all of the species in an ecosystem instead of focusing on a single species.

Science in the News: Critical Thinking Worksheets Answer Key

Segment 21
Fusion for Power?
1. Answers will vary. Sample answer: The justification is that fusion is safer and cleaner than conventional energy sources. It is also a nearly limitless source of energy. As natural fuels, such as oil, become more scarce, power from fusion could be an extremely valuable resource.
2. Answers will vary. Sample answer: Scientists want to keep the particles from touching the sides because the particles are very hot and are moving very fast. If they touch the metal walls, they might escape into the air.
3. Unlike traditional power plants or nuclear power plants, plants that operate on fusion create usable energy without causing acid rain, forming greenhouse gases, or producing radioactive waste.
4. Fusion is referred to as a limitless energy source because it uses sea water and lithium, both of which are very abundant on Earth.

Segment 22
Geothermal Energy
1. A geothermal energy system reduces the amount homeowners must pay to heat and cool their homes.
2. Answers will vary. Sample answer: No; a backup energy source is not necessary because ground temperatures stay fairly constant all year.
3. Geothermal energy reduces the need for electricity, thereby reducing the pollution created by the production of electricity, which often involves fossil fuels.
4. Answers will vary. Sample answer: Geothermal energy may not supply enough power for all needs. Also, buildings located in places where it would be difficult to dig may not benefit from geothermal energy.

Segment 23
Talking Trash
1. Answers will vary. Students' lists should include common items, such as plastic packaging, as well as perishables, such as food leftovers. Students should realize that a large percentage of what they throw away could be recycled or reused in some way.
2. Answers will vary. Sample answer: All communities don't have a recycling program because of the short-term costs involved in starting the program as well as concerns about each community's willingness to participate.
3. Answers will vary. Answers should include information about what raw materials were used to make and package the product, how the product was used, and if it was recycled or went to a landfill.
4. Answers will vary. Sample answer: A banana peel can be composted to make fertilizer. A used pair of shoes can be donated to a center for the needy. The paper in most books can be recycled to make new paper products.

Segment 24
Treating Toxic Waste
1. Many of the chemicals are no longer harmful once they are changed back into their simplest state. Also, many of the chemicals can be reused.
2. Radioactive waste has not been tested with molten metal technology.
3. The Department of Energy sees this technology as a possible way to dispose of radioactive wastes.
4. This new technology is both cost efficient and energy efficient compared with existing methods of treatment and storage. All of the byproducts can be reused or sold to industries.

Science in the News: Critical Thinking Worksheets Answer Key

Segment 25
Critical Masses
1. Answers will vary. Sample answer: The average life expectancy in extremely overcrowded cities is probably lower due to disease and pollution.
2. Answers will vary. Sample answer: When a population becomes too large, many people are forced into poverty by a lack of work, services, and housing. These shortages promote disease and other problems. Overpopulation can also overwhelm systems that control pollution.
3. Answers will vary. Sample answer: People move to cities because there is a perception that there are more jobs and services available in cities than in rural areas.
4. Rapid population growth is occurring mostly in developing countries, although many developed countries also have problems with overcrowding in cities.

Segment 26
Declining Birthrates
1. The high cost of living discourages people from having large families. Also, services, such as children's day care, are scarce and expensive.
2. The Japanese government is concerned about the declining population because its economy relies on a large number of workers.
3. Answers will vary. Sample answer: The Japanese government could slow its population decline by building more housing to relieve crowding and by lowering taxes to reduce the cost of living. The Japanese government could also encourage couples to have more children by offering housing and educational benefits to growing families.
4. Answers will vary. Sample answer: People in industrialized nations might have fewer children due to the availability of family planning and education. Many people in developing countries do not have access to these services.

Segment 27
A Climate Conference
1. The attendee means that many countries say they want to help reduce greenhouse gases and other environmental pollutants, but they are unwilling to take the political and financial steps to make that happen.
2. Answers will vary. Sample answer: It may be considered extreme to say that there is no evidence that global warming is caused by industry emissions.
3. Answers will vary. Sample answer: Developing nations have a greater need for immediate financial improvements and may not have as many concerns with the long-term effects of pollution and global warming.
4. Answers will vary. Sample answer: The main advantage of a binding environmental agreement is that countries are obligated to complete the agreement. However, it is more difficult to convince countries to make binding agreements.

Segment 28
Something Worth Saving
1. The tropical forest is referred to as a "living laboratory" because of the large number of plant and animal species that interact within it.
2. Answers will vary. Sample answer: Tropical species may have components that could be used in medicines or may contain chemicals that could be used in insecticides.
3. Answers will vary. Sample answer: A biological prospector might be interested in collecting insect specimens, which could contain clues about why some plants are eaten by bugs and other plants are not.
4. a. Answers will vary. Sample answer: Preserving the rain forest helps to maintain global biodiversity, which in turn keeps the food chain healthy and intact.
   b. Answers will vary. Sample answer: Scientists contribute to this effort by studying and identifying threatened species.